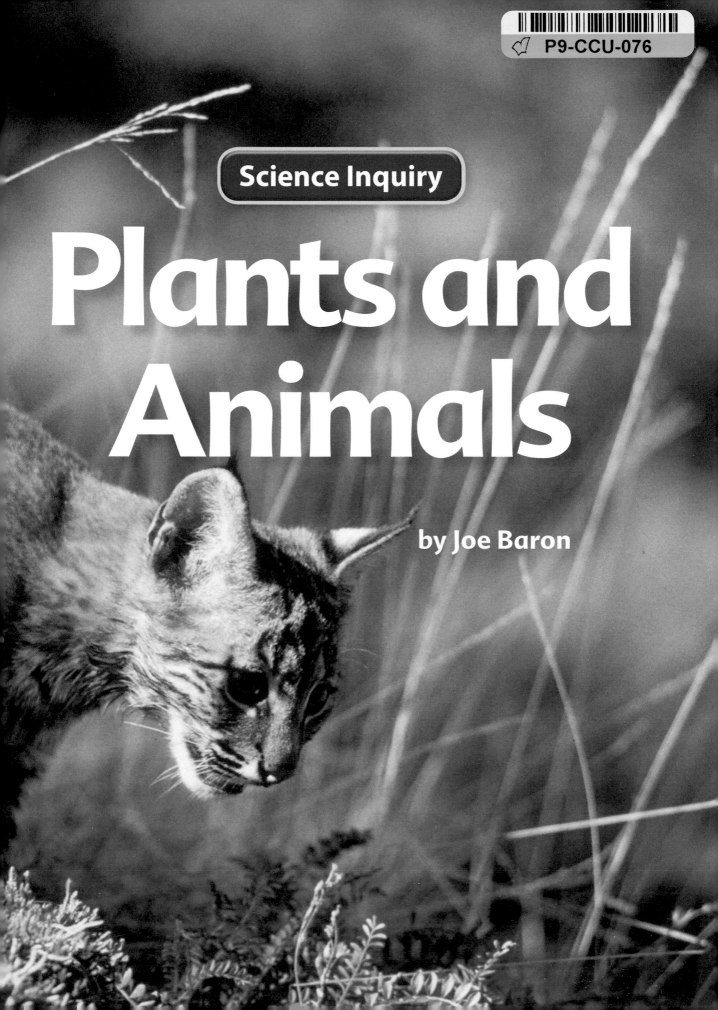

Science Inquiry

Plants and Animals

by Joe Baron

Science Inquiry

Science in a Snap!

Explore Activity

▶ **Question:** How are the seeds of sunflower plants alike
and different?

Directed Inquiry

▶ **Question:** How do the parts of two kinds of
plants compare?

Directed Inquiry

▶ **Question:** How can you sort some animals
by their body coverings?

Science in a Snap!

Think about the parts of a plant. What does your plant part do for the plant? Write a riddle about one part. Read your riddle out loud. Have a partner guess the answer.

I am a part of a plant.
I make seeds.
What am I?

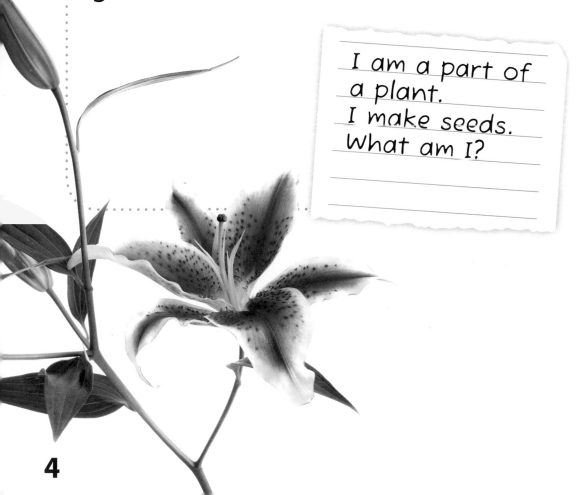

Animal Antics

Think of a body part that an animal uses to eat, drink, or move. Act out how the animal uses that body part. Have classmates guess what part you are and what you are doing.

Be a Seed Detective

Open a bean seed. Use a hand lens to observe what is inside. Draw what you see.

Investigate Plants

Question How are the seeds of sunflower plants alike and different?

Science Process Vocabulary

observe verb

When you **observe,** you use your senses to learn about an object or event.

share verb

When you **share** results, you tell or show what you have learned.

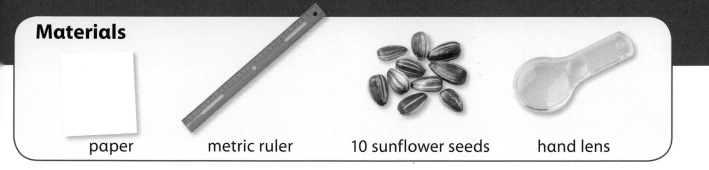

paper   metric ruler   10 sunflower seeds   hand lens

What to Do

1 Draw a line on a sheet of paper. Make the line 15 centimeters long.

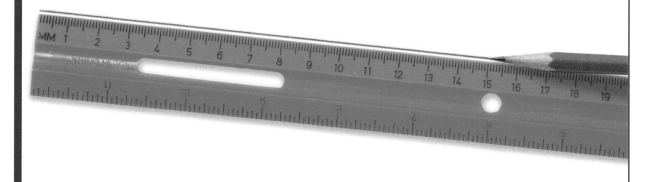

2 Place 10 sunflower seeds on the line. Look at the shape of each seed. Do all seeds have the same shape? Record what you **observe** in your science notebook.

3 Observe the sunflower seeds. Are they all the same size? Record your observations in your science notebook.

4 Use the hand lens to look at each sunflower seed. **Count** the number of stripes on each seed. Record the **data** in your science notebook.

Record

Write in your science notebook. Use a table like this one to record the number of stripes you observe.

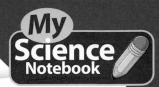

Sunflower seed	Number of stripes
1	
2	
3	

Share Results

1. Tell how the seeds are like each other.

The seeds have the same _____.

2. Tell how the seeds are different from each other.

The seeds have different _____ and different _____.

9

Investigate Plant Parts

> **Question** How do the parts of two kinds of plants compare?

Science Process Vocabulary

observe verb

You can use a hand lens to help you **observe** objects.

compare verb

You **compare** plants when you look at them to see how they are alike or different.

Materials

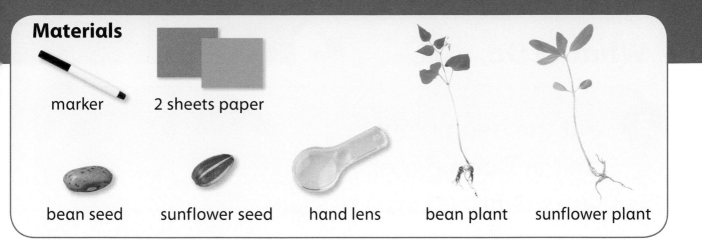

marker 2 sheets paper

bean seed sunflower seed hand lens bean plant sunflower plant

What to Do

1 Label 2 sheets of paper.

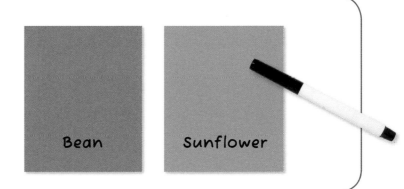

Bean Sunflower

2 Put the correct seed on each paper. Use the hand lens to **observe** each seed. Record what you observe in your science notebook.

Sunflower

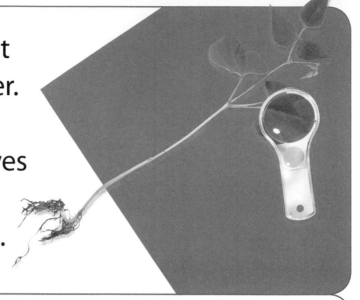

3 Put the bean plant on the **Bean** paper. Use the hand lens to observe its leaves and stem. Record your observations.

4 Observe the roots. Record your observations.

5 Repeat steps 3 and 4 with the sunflower plant.

6 Observe the pictures of the flowers. Record your observations.

▲ Bean

▲ Sunflower

Record

Write or draw in your science notebook. Use a table like this one.

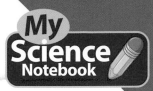

	Bean	Sunflower
Seed		
Leaves		

Explain and Conclude

1. **Compare.** How are the plant parts alike?

2. How are the plant parts different?

Think of Another Question

What else would you like to find out about how plants are alike and different?

Investigate Animal Traits

Question How can you sort some animals by their body coverings?

Science Process Vocabulary

sort verb

When you **sort,** you put things in groups.

Fish

Mammals

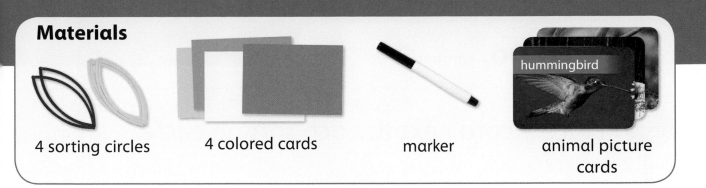

4 sorting circles   4 colored cards   marker   animal picture cards

What to Do

1 Unfold your sorting circles.

2 Label 4 cards.

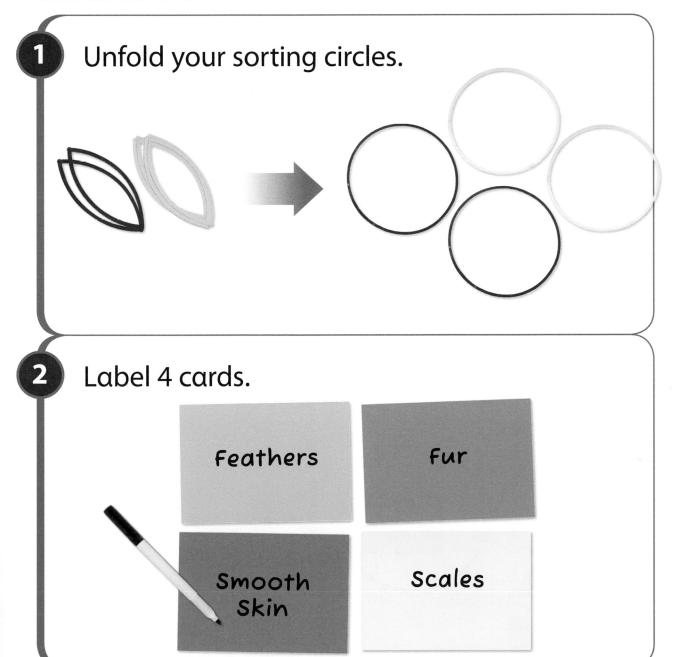

Feathers

Fur

Smooth Skin

Scales

3 Place 1 card next to each sorting circle.

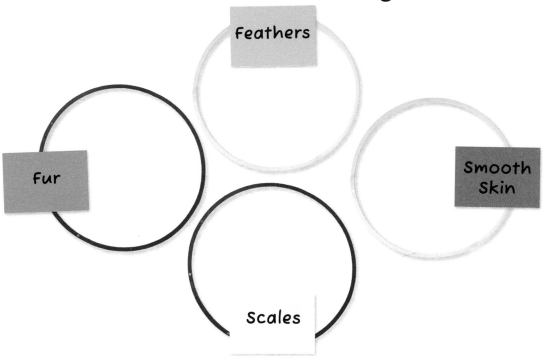

Feathers

Fur

Smooth Skin

Scales

4 **Sort** the animals by the kind of body coverings they have.

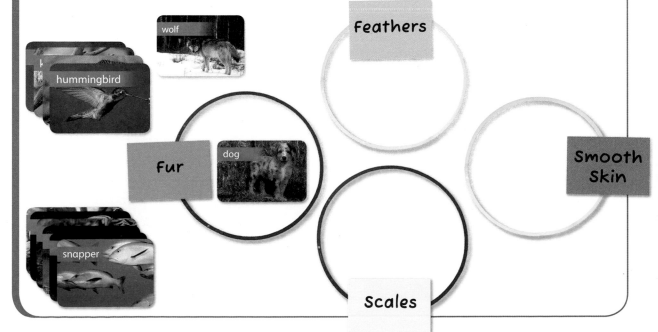

wolf

hummingbird

Feathers

Fur

dog

Smooth Skin

snapper

Scales

Record

Write or draw in your science notebook. Use a table like this one.

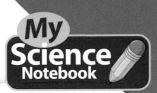

Fur	Feathers	Smooth Skin	Scales
dog			

Explain and Conclude

1. Explain how the body coverings of fish and birds are different.

2. In which group did you put each animal?

Think of Another Question

What else would you like to find out about the body coverings of animals? What could you do to answer this new question?

Math in Science

Graphs

You can organize the data you collect in a graph. John went to the zoo. He saw that animals have different body coverings. He recorded his observations in a table.

Body Covering	Number of Animals
Fur	I I I I
Feathers	I I
Scales	I I I
Smooth Skin	I

John used the information in his table to make a graph. Look at the bar for **Feathers.** It stops at the line that is labeled **2.** That means that John saw 2 animals that had feathers as a body covering.

The **Scales** bar is longer than the **Feathers** bar. John saw more animals with scales than animals with feathers.

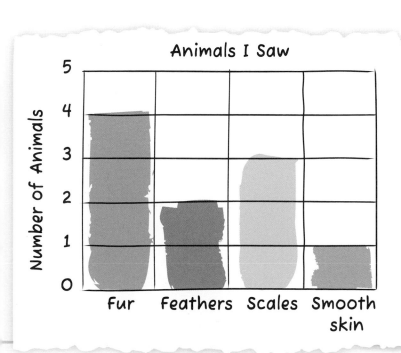

Animals I Saw

Number of Animals

5
4
3
2
1
0

Fur   Feathers   Scales   Smooth skin

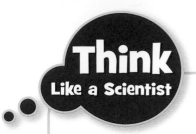
A graph shows information in a way that is easy to understand and use. Make a graph to organize data you collect in an investigation.

▶ What Did You Find Out?

1. How many animals with scales did John see?

2. Which kind of body covering did John see most?

3. Did John see more animals with scales or with fur? How do you know?

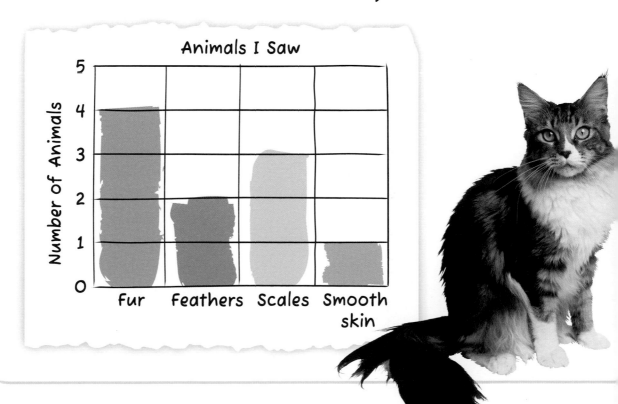

Animals I Saw

Make and Use a Graph

1. Ask 10 students to name a favorite kind of pet. Record the answers in a chart. Decide what kind of body covering each pet has.

2. Make a graph.

   - Choose 4 kinds of body coverings.

   - Count the number of pets with each kind of body covering.

   - Color in boxes to show the number of animals with each body covering.

3. Share your graph with a partner and ask questions about it.

Investigate Bird Beaks

Question Which food will a model bird beak best pick up?

Science Process Vocabulary

predict verb

When you **predict,** you tell what you think will happen.

If a bird's beak is small and short, then it will be able to pick up seeds.

fair test

An investigation is a **fair test** when you change only one thing and keep everything else the same.

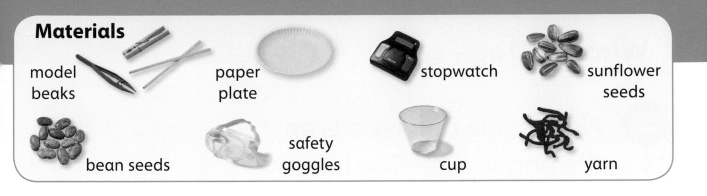

Materials

model beaks • paper plate • stopwatch • sunflower seeds • bean seeds • safety goggles • cup • yarn

Do a Fair Test

Write your plan in your science notebook.

Make a Prediction

In this investigation, you will test a model bird beak to see what food it picks up best. **Predict.** Which food will your model best pick up?

Plan a Fair Test

What one thing will you change?
What will you observe or measure?
What will you keep the same?

What to Do

1. Choose a **model** bird beak to test.

2 Place some bean seeds on the paper plate. The bean seeds stand for beetles that birds might eat.

3 Put on your safety goggles. Have a partner use a stopwatch to time you for 30 seconds. Try to pick up as many bean seeds as you can with your model bird beak. Record your **data** in your science notebook.

Put the food you ▶ **pick up in the cup.**

4 Birds also eat seeds. Do steps 2 and 3 with the sunflower seeds.

5 Do steps 2 and 3 with the pieces of yarn. The yarn stands for worms.

Record

Write in your science notebook. Use a table like this one.

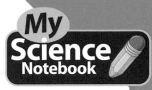

Food Pieces Picked Up with _____		
Bean seeds (beetles)	Sunflower seeds (seeds)	Yarn (worms)

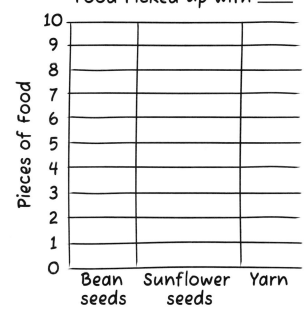

Food Picked Up with _____

Make a graph.

Explain and Conclude

1. Does the **data** in your graph support your **prediction?** Tell why.

2. **Compare** the data from all groups. Describe any **patterns** you see.

Think of Another Question

What else would you like to find out about bird beaks?

Investigate How an Animal Changes

Question What changes can you observe in a darkling beetle during its life cycle?

Science Process Vocabulary

observe verb

When you **observe** something, you use your senses to learn about it.

infer verb

When you **infer,** you use what you know and what you see to explain something.

I can infer that the larva is changing during the pupa stage.

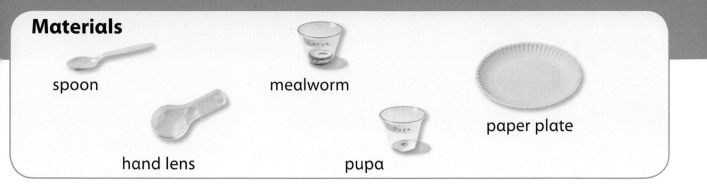

Materials

spoon

mealworm

paper plate

hand lens

pupa

What to Do

1 Use a spoon to put the mealworm on the paper plate. Handle the mealworm gently. **Observe** the mealworm with the hand lens. Record your observations.

A mealworm is ▶ the larva of a darkling beetle.

2 Choose a behavior. Decide how to test the behavior with your mealworm. Test the behavior. Record how your mealworm acts.

Behaviors
- Does it move when you touch it with the spoon?
- Does it move toward a moist paper towel?
- Does it move toward food?

3 Put the mealworm back in the cup.

4 Use the spoon to put the pupa on the paper plate. Observe the pupa. Test the behavior. Record your observations.

5 Put the pupa back in the cup. Observe the pupa every day until it changes to an adult darkling beetle. Record your observations.

Record

Write and draw in your science notebook. Use a table like this one.

	Obervations
Mealworm (larva)	
Pupa	

Explain and Compare

1. What behavior did you test? What can you **infer** from the behavior?

2. **Compare** the darkling beetle larva and pupa.

3. How is the mealworm different from the adult darkling beetle? How is the pupa different from the adult darkling beetle?

Think of Another Question

What else would you like to find out about the life cycle of the darkling beetle?

Adult darkling beetle ▶

Do Your Own Investigation

Question **Choose a question, or make up one of your own to do your investigation.**

- Will the bean plants grown from seeds in the same package be alike?

- How are mealworms alike and different?

- How does a radish plant change through its life cycle?

Science Process Vocabulary

data noun

When you gather **data,** you collect information.

Plant 1   Plant 2   Plant 3

2 cm   3 cm   2 cm

Open Inquiry Checklist

**Here is a checklist you can use
when you investigate.**

- ☐ Choose a **question** or make up one of
  your own.

- ☐ Gather the materials you will use.

- ☐ Tell what you **predict.**

- ☐ Make a **plan** for your investigation.

- ☐ Carry out your **plan.**

- ☐ Collect and record **data.** Look for **patterns** in
  your data.

- ☐ Explain and **share** your results.

- ☐ Tell what you **conclude.**

- ☐ Think of another
  question.

Think Like a Scientist

How Scientists Work

Gathering Information

Scientists can gather information in many ways.

Read books. Suppose you wanted to find out about gorillas. You might read books about them. Scientists also use books to find information.

Ask an expert. Scientists might also ask an expert. An expert is someone who knows a lot about a certain thing.

Make observations. Scientists make observations to learn about things. Many scientists have observed gorillas living in Africa. They have learned how gorillas live and when they sleep.

Watch a video. Not all scientists can go to Africa. But they still can observe gorillas. They can watch a video that shows gorillas and where they live.

▼ Michael McRae traveled to Africa to study gorillas.

Do investigations. Scientists also do investigations to find information. Scientists might investigate to find out what kinds of food gorillas eat.

▶ **What Did You Find Out?**

1. What are 5 ways scientists gather information?

2. What could you find out about an animal by observing it?

Gather Information

1. Choose an animal you would like to know more about.

   - Write a question about the animal.

   - Choose 2 ways you will look for answers to the question.

   - Research the question.

2. Tell how each way you looked for information helped you answer the question.

Featured Photos

Cover: zebra caterpillar moth *(Melanchra picta* larva) on buttercup plant, North Carolina

Title page: bobcat kitten

page 9: sunflowers

page. 13: primroses

page 17: laboratory rat

page 18: green snake coiled around a branch; 2 carps swimming; brown bear family in stream, McNeil River State Game Sanctuary, Alaska

page 19: 2 saw-whet owls perched on branch; common frog *(Rana temporaria)*, England

page 20: domestic cat

page 21: boy with iguana; goldfish

page 31: radish plants in garden soil

page 32: a gorilla and rehabilitator, Lefini Faunal Reserve, Congo

page 34: group of mountain gorillas *(Gorilla gorilla berengel)*

page 35: cattle egret on a burchell's zebra, Amboseli National Park, Kenya

inside back cover: Lined seahorse *(Hippocampus hudsonius)*

ACKNOWLEDGMENTS
Grateful acknowledgment is given to the authors, artists, photographers, museums, publishers, and agents for permission to reprint copyrighted material. Every effort has been made to secure the appropriate permission. If any omissions have been made or if corrections are required, please contact the Publisher.

PHOTOGRAPHIC CREDITS
set-up photography: Andrew Northrup; Cover (bg) Creatas/Jupiterimages; Title (bg) Corel; 4 rgbstudio/Alamy Images; 5 (t) Anup Shah/Photodisc/Alamy Images; 6 (t) Terry Why/Jupiterimages, (c) LWA-Sharie Kennedy/Corbis, (b) Comstock Images/Jupiterimages/Alamy Images; 9 Petr Vaclavek/Shutterstock; 10 (t) Robert Madden/National Geographic Image Collection, (c, b) Fancy/Veer/Corbis; 12 (bl) Maxine Adcock/Photo Researchers, Inc., (br) Dallas and John Heaton/ SCPhotos/Alamy Images; 13 PhotoDisc/Getty Images; 14 (t) PhotoDisc/Getty Images; 17 Bill Gallery - Doctor Stock/Science Faction/Corbis; 18 (l) MetaTools, (c) Juniors Bildarchiv/Alamy Images, (r) Jenny E. Ross/Corbis; 19 (t) Ron Austing; Frank Lane Picture Agency/Corbis, (b) Robert Maier/Animals Animals, 20 G.K. & Vikki Hart/Photodisc/Getty Images; 21 (l) Corbis Premium RF/Alamy Images, (r) Photodisc/Getty Images; 22 (t) Corel, (b) LWA-Sharie Kennedy/Corbis; 26 (t) Creatas/Jupiterimages, (b) Roy Morsch/Corbis; 29 G.K. & Vikki Hart/Photodisc/ Getty Images; 30 (t) Kris Timken/Getty Images; 31 Visuals Unlimited/Corbis; 32, 33 Michael Nichols/National Geographic Image Collection; 32-33 (bg) George Steinmetz/Corbis; 34 Digital Vision/Getty Images; 35 Tim Wright/Corbis; Inside Back Cover Digital Vision/Getty Images.

ILLUSTRATOR CREDITS Amy Loeffler

Neither the Publisher nor the authors shall be liable for any damage that may be caused or sustained or result from conducting any of the activities in this publication without specifically following instructions, undertaking the activities without proper supervision, or failing to comply with the cautions contained herein.

PROGRAM AUTHORS
Judith Sweeney Lederman, Ph.D., Director of Teacher Education and Associate Professor of Science Education, Department of Mathematics and Science Education, Illinois Institute of Technology, Chicago, Illinois; Randy Bell, Ph.D., Associate Professor of Science Education, University of Virginia, Charlottesville, Virginia; Malcolm B. Butler, Ph.D., Associate Professor of Science Education, University of South Florida, St. Petersburg, Florida; Kathy Cabe Trundle, Ph.D., Associate Professor of Early Childhood Science Education, The Ohio State University, Columbus, Ohio; Nell K. Duke, Ed.D., Co-Director of the Literacy Achievement Research Center and Professor of Teacher Education and Educational Psychology, Michigan State University, East Lansing, Michigan; David W. Moore, Ph.D., Professor of Education, College of Teacher Education and Leadership, Arizona State University, Tempe, Arizona

THE NATIONAL GEOGRAPHIC SOCIETY
John M. Fahey, Jr., President & Chief Executive Officer
Gilbert M. Grosvenor, Chairman of the Board

National Geographic School Publishing
Hampton-Brown
www.NGSP.com

Printed in the USA.
RR Donnelley, Johnson City, TN

ISBN: 978-0-7362-6230-9

10 11 12 13 14 15 16 17

10 9 8 7 6 5 4 3 2